BODY SYSTEMS:
Skeletal
and
Muscular

BY GARY RUSHWORTH

Table of Contents

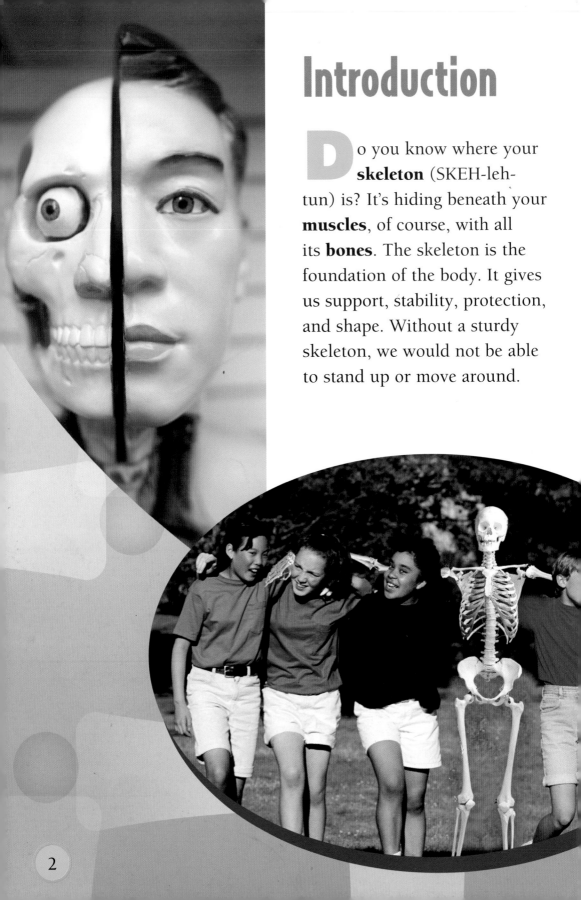

Introduction

Do you know where your **skeleton** (SKEH-leh-tun) is? It's hiding beneath your **muscles**, of course, with all its **bones**. The skeleton is the foundation of the body. It gives us support, stability, protection, and shape. Without a sturdy skeleton, we would not be able to stand up or move around.

The skeleton lets us stand upright. It helps hold our organs in place and protects them, too. It makes movement possible, but not without help. That is what muscles are for. But they can't do that without the skeleton. Both must work together.

Muscles are attached to the bones and connected to the brain by nerves. When you want to run or walk, stand or sit, or jump up and down, the brain tells your muscles to tighten up or relax in a certain way. The next thing you know, you are walking, running, jumping, swimming, or sitting down for a rest.

In this book, you will learn a lot about the skeleton and muscles. You will learn how bones are made and how they attach together to form the skeleton. You will learn about muscles, some of their names, and how they work with bones and **joints** (JOINTS). Finally, you will learn how those parts all work together so we can walk, run, and move around.

As you read this book, you use your skeleton and muscles. Your skeleton provides support, and your muscles make your heart beat, your lungs breathe, and your brain work. Muscles help your eyes move as you read about these two important systems.

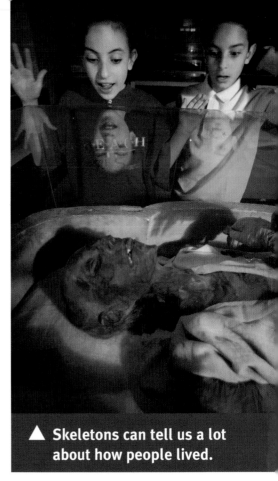

▲ **Skeletons can tell us a lot about how people lived.**

Careers in Science

Archaeologists study the remains of people from ancient civilizations to see how they lived and worked.

4

Chapter 1

The Skeleton
(All Put Together)

The skeleton, all put together, is one of the best ways to start looking at bones and muscles. Why? Because the skeleton is the framework of the body. It holds our organs in place and protects them. The skeleton provides attachments for our muscles to make movement possible.

When you look at a complete skeleton, you can see all the bones in the human body. You can also see how bones attach at joints. Take a look at the skeleton on page 6. What is the first thing you notice? There are no wrong answers here. Think about what you notice first and why.

It's a Fact

The human body has 206 bones! There are about 100 bones in the hands and feet alone. Babies are born with about 300 bones. But you don't lose bones as you get older. Some bones, like the ones in your skull, join together as you get older.

The Human Skeleton

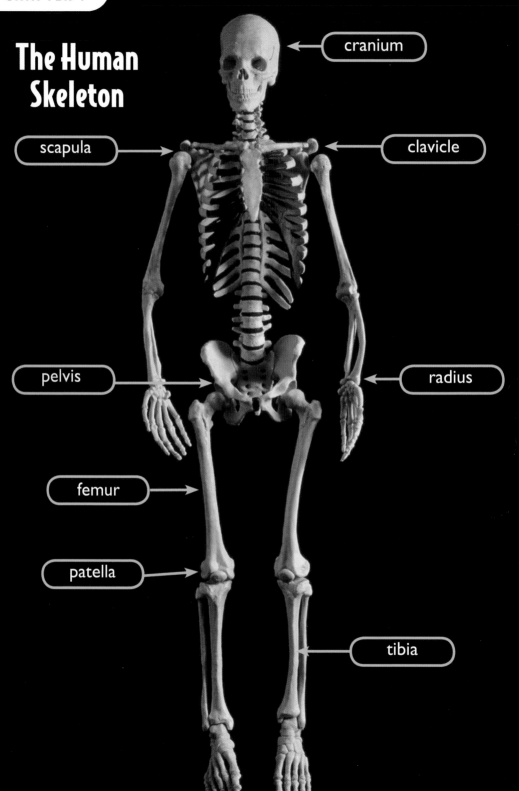

cranium

scapula

clavicle

pelvis

radius

femur

patella

tibia

The Skeleton
(Taken Apart)

Most people see the **skull** first. It is a part of the skeleton that is easily recognized and identified. It protects the brain, fixes the position of the eyes and ears, and gives shape to the skin and other tissues of the face.

It's a Fact

There are twenty-nine bones in the skull, including the jaw and face. When we are born, the bones of the skull are separate and flexible. As we grow older, the bones grow together and form a solid skull.

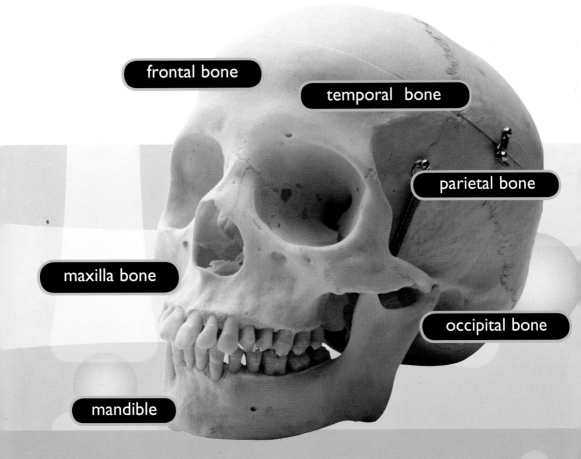

frontal bone

temporal bone

parietal bone

maxilla bone

occipital bone

mandible

▲ The human skull has more than twenty bones.

Bone is a rigid, tough connective tissue that makes up the skeleton. Not all of bone is hard, though. Most bones have a soft center. This is called **marrow**.

marrow

▲ The soft center of your bones is called marrow.

Bone marrow is very important. It produces white blood cells that help fight infections and disease. It also produces red blood cells. These cells carry oxygen to all body cells to help the cells function.

Bone marrow also produces platelets, which aid in blood clotting.

Bones also contain huge amounts of minerals. These minerals are important to many body processes.

There are 206 bones in the human body. There are twenty-nine bones in the skull and 100 in the hands and feet. Bones attach to other bones and muscles to allow for movement and support.

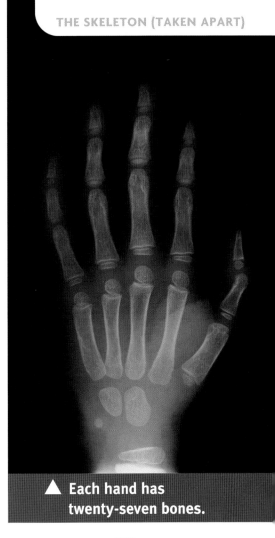

▲ Each hand has twenty-seven bones.

Careers in Science

Physical therapists are health care professionals that help patients with broken bones, and damaged joints and muscles.

The skull sits on top of and attaches to the **spinal column**. The spinal column is the central support of the entire skeleton. It not only supports, but it allows us to bend, twist, and turn our bodies.

The major part of the skull (the part that covers the brain) is the **cranium**.

The cranium covers the brain with a very hard shell. This shell is necessary to prevent damage to the brain should the head come in contact with another hard surface, such as might happen in falling.

▲ **This girl's helmet protects her head.**

The cranium is very strong, but it can still be damaged. So, you must protect it. Always wear a helmet when you skateboard, ride your bike, or play football.

If you look closely at the skull, you will see that the lower jaw is connected to the cranium just in front of the ears. The lower jaw is called the mandible. It moves and contains teeth. Together, the cranium, the mandible, and other facial bones form the skull.

▲ Whenever you play a sport, protect your skull. ▲

As you read, the spinal column is the major support structure of the skeleton. It holds the head up, supports the chest, and attaches to the pelvis.

The spinal ▲ ▶ column is the center post of the skeleton and allows us to twist and bend.

The spinal column is actually a group of bones called **vertebrae** that are connected. Vertebrae are individual bones that fit together like building blocks. Vertebrae are stacked on top of each other.

Between the vertebrae and separating them is a tough material called a **disk**. The disks act like shock absorbers, keep the vertebrae from rubbing together and allow the back to move.

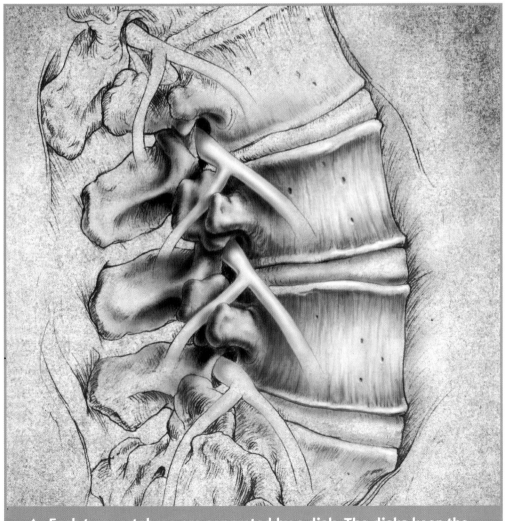

▲ Each two vertebrae are separated by a disk. The disks keep the vertebrae from rubbing against each other and causing pain.

13

Disks are made of **cartilage** (KAR-tih-lij). Cartilage is a kind of tissue called connective tissue. It is called connective tissue because it holds tissues together. Cartilage is spongy, smooth, and flexible. It covers bone ends and joints. It is also the type of support tissue that makes up the ears and nose.

Through the center and along the entire length of the spinal column is the spinal cord, which lies in the spinal canal. The spinal cord is the main nerve that connects the brain to the rest of the body.

The entire spinal cord is bathed in a clear fluid called spinal fluid. Spinal fluid surrounds the entire spinal cord and protects the spinal cord from damage by the disks. There are also blood vessels and many smaller nerves located within the spinal canal.

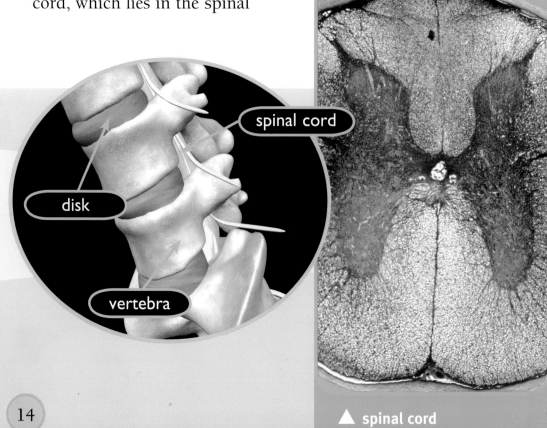

spinal cord

disk

vertebra

▲ spinal cord

Attached to the spinal column on the upper end is the rib cage. The rib cage forms the bones of the chest. The rib cage covers and protects the heart and lungs from injuries.

The ribs join in the center of the chest to the sternum (STER-num). The sternum is part-bone and part-cartilage. The cartilage helps with flexibility and the bone makes it strong.

At the bottom of the spinal column is the pelvis. The pelvis gives the legs something to attach to. It forms our hips. The pelvis protects the lower abdomen, as well, and gives us something to sit on. Muscles provide the padding. Muscles also help us stand up and sit down.

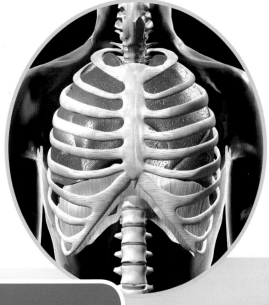

▲ rib cage

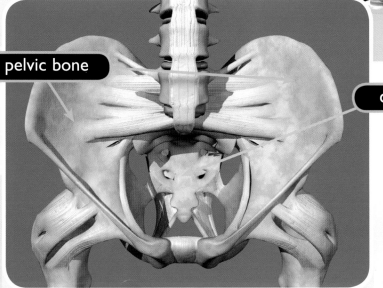

pelvic bone

coccyx

15

Attached to the rib cage on the top are the collar bones, or clavicles. These bones together with two triangular-shaped bones on the back, called the scapula, form the bones of the shoulders. Your arms attach where the rib cage, scapula, and clavicles meet.

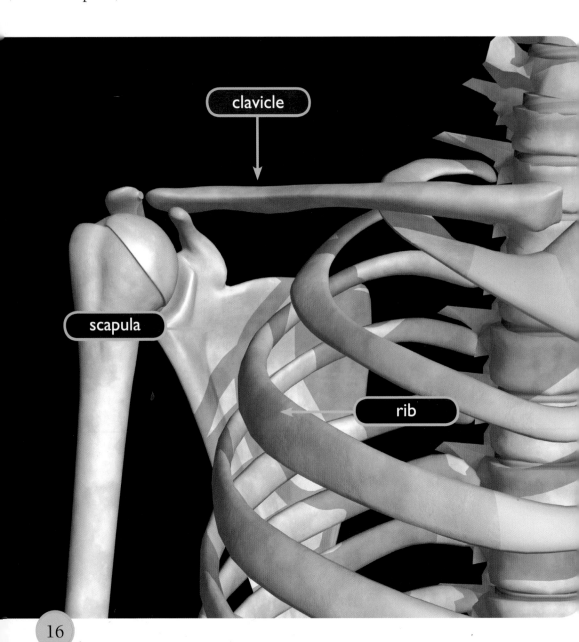

clavicle

scapula

rib

Joints: Bone to Bone

You will notice that your arms and legs are composed of several bones. Your hands attach to your lower arms and your feet attach to your lower legs. The point where each bone meets another bone is called a joint. Joints allow bones to move without damaging each other. They keep the bones far enough apart so they do not rub together as they move while holding the bones in place. There are five kinds of joints.

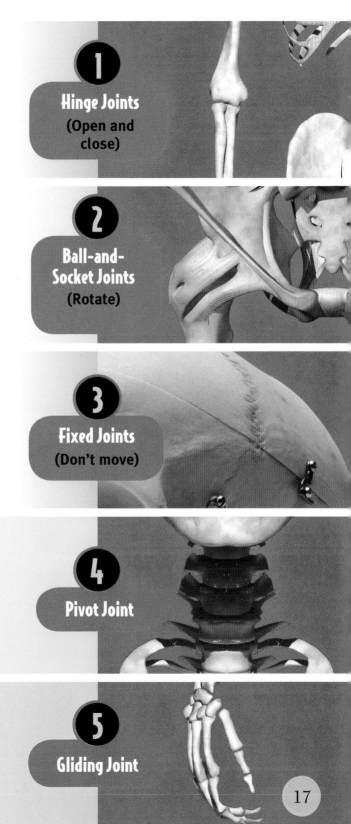

1 Hinge Joints (Open and close)

2 Ball-and-Socket Joints (Rotate)

3 Fixed Joints (Don't move)

4 Pivot Joint

5 Gliding Joint

17

Joints allow us to reach, pull, walk, and kick our arms and legs (hinge joint); rotate our hips and shoulders (ball-and-socket joint); and hold two or more bones together, like the bones of our skull (fixed joint).

1 ball-and-socket joint

2 hinge joint

Chapter 3

The Muscles
(And Ligaments and Tendons)

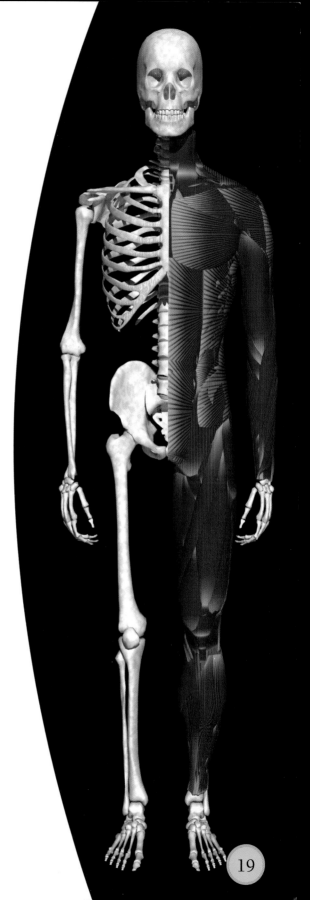

The skeleton makes up the framework and support for the body. The skeleton provides protection for our organs, especially the brain, lungs, and heart. We know it is made of bones and cartilage. We know each bone is connected to another by a joint. We know the skeleton also makes movement possible, and that it needs muscles for movement to happen.

A muscle is a part of the body that can contract (tighten up) and extend (relax) in a specific order to allow movement to occur. The tightening and relaxing of the muscles of the body make the body work and do the things it needs to do.

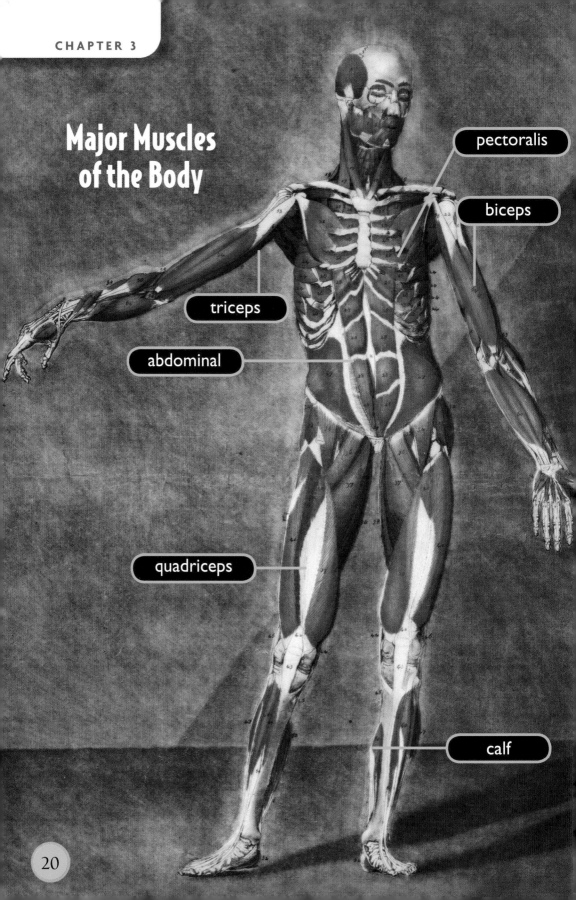

Major Muscles of the Body

pectoralis

biceps

triceps

abdominal

quadriceps

calf

There are three kinds of muscles: smooth, skeletal, and cardiac. Although they are all muscles, they work in different ways.

Smooth muscles are muscles that we do not have control over. These are the muscles found in the stomach, intestines, lungs, and other internal organs. Smooth muscles make up the walls of many blood vessels. They are called "involuntary" muscles because we can't control what they do.

Skeletal muscles are usually attached to bones. They are muscles that we can control. They are the large muscles of the arms and legs, abdomen, and back. They are also the smaller muscles of the fingers, toes, eyes, and mouth. Skeletal muscles are voluntary muscles. They control walking, running, and other activities.

Cardiac muscles are muscles found only in the heart. They cause our hearts to pump blood through the body. These muscles, like smooth muscles, work continuously: twenty-four hours a day, seven day a week. They truly keep us alive.

Point

Think About It

Review the three types of muscles discussed. Think of things your body does every day. See if you can tell which type of muscle performs the function.

Muscles work in two ways. They tighten up, or contract, and they relax, or extend. This alternate contraction and relaxation is what causes us to move. But it takes time for this tightening and relaxing to occur.

The relaxation period between when a muscle contracts and relaxes and when it is able to contract again is called the recovery time. As muscles become tired, this time gets longer and longer. This is called muscle fatigue.

It's a Fact

There are over 600 kinds of muscles that we know of. Some are tiny, like the muscles that give us goose bumps when we are cold. Some are large, like the muscles of your legs.

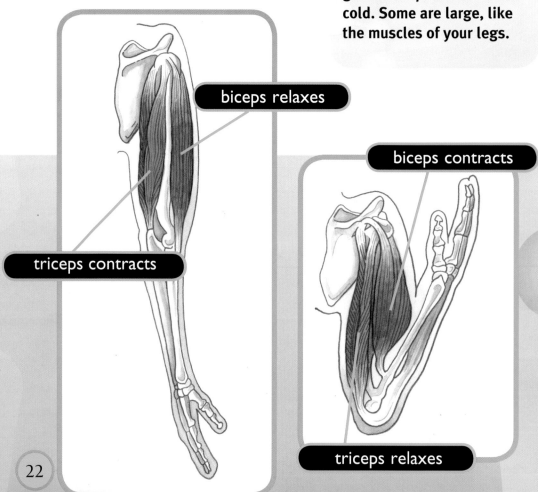

biceps relaxes

biceps contracts

triceps contracts

triceps relaxes

Muscle fatigue is a big problem for the body. But the body is a great problem solver. When a muscle contracts, it expends a great deal of energy. That happens pretty quickly. But recovery is slow. So most major muscles work in pairs. For instance, the muscles in the front of your arms and legs contract to move your arms and legs forward. Once that movement occurs, those muscles need time to recover.

Without a recovery period, you would extend or put your leg forward to walk and you would get stuck until the muscle recovered. The body deals with this by having the muscles in the back of your leg contract, allowing the muscles in front to recover.

The same thing is happening in the other leg, except that while one leg is moving forward the other is moving back.

Skeletal muscles help us do many things. They help us push, pull, lift, throw, and smile. They help us to chew, swallow, and digest our food, and play video games and sports. Muscles are highly organized and work well together and with the bones of the skeleton. But before the muscles and bones can work together, they must attach to each other.

Skeletal muscles attach directly to bones by **tendons**. A tendon is a tough, fibrous tissue cord that is part of the muscle. This cord attaches itself permanently to a part of the bone, holding one end of the muscle to the bone. By pulling on the bones, tendons produce movement.

To allow movement at a joint, both bones must be connected. A tough, fibrous band of tissue called a **ligament** attaches to the membranes that surround the bones and holds the bones together.

▲ Our muscles help us play sports better.

The point where a ligament attaches to a bone is called the origin. The point where a tendon attaches to a bone is called the insertion. When a muscle contracts, it tightens along its length, or shortens. This pulls on the tendon and ligament, causing them to pull on the bone they are attached to.

Because the bones are connected by a joint, this contraction causes the bones to move. The kind of joint (hinge, ball-and-socket, or gliding) determines the type of movement.

✔ Point

Reread

Think about the function of each joint while rereading pages 17 and 18. For now, just consider hinge, ball-and-socket, and fixed joints. Try to think of something you do every day, like walking. What type of joints are involved and why?

tendon

ligament

tendon

Cardiac muscles are found only in the heart. They pump blood from the heart to all parts of the body.

Cardiac muscle looks like skeletal muscle in that it appears to have bands or stripes. However, unlike skeletal muscle, cardiac muscle is involuntary muscle.

Because of what cardiac muscle does, it must be able to work automatically, day and night. It is always pumping blood throughout the body.

It has to be a strong and tough muscle to do that.

The heart is a muscular pump. Actually, it is four pumps in one, all working together. One side of the heart pumps blood to the lungs to get rid of carbon dioxide and pick up oxygen, returning it to the heart. The other side of the heart pumps blood to the body. The movement of blood through the body is called circulation.

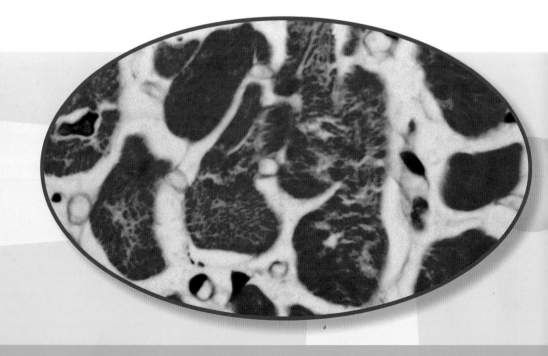

▲ cardiac muscle

Suppose a drop of blood is in the lungs and has just picked up oxygen to bring to the cells. The right side of the heart contracts, or squeezes, and pushes the blood out of the lungs and into the left side of the heart.

The left side of the heart pumps the blood containing oxygen to all the cells of the body. With each heartbeat, the blood is pushed through the arteries and capillaries to all the cells.

When the blood reaches the cells, oxygen is released and carbon dioxide is absorbed. The carbon dioxide will be carried to the lungs and eliminated. This occurs with every heartbeat throughout your entire life.

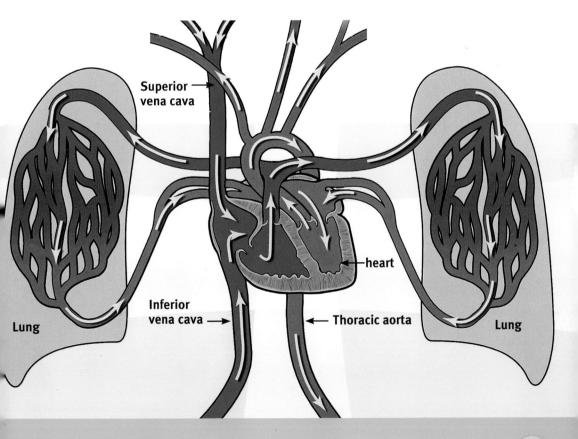

▲ blood pumping out to the body

The Joy of Moving Around

Forward and backward, side to side, and up and down. Movement is a wonderful thing. Our skeleton and muscles make this possible, with help from our brain, of course. We can do the things we need to do, get from place to place, and experience so much. There is no limit to the things we can learn to do and the things we can see.

Exercise makes your muscles fitter, bigger, and stronger. This makes you feel better and able to do things more easily, like walking or running.

How much walking and running you can do depends on how much you exercise and the kinds of exercise you do. So walk a mile, run some laps, ride a bike, play baseball or soccer—or anything you can think of to keep moving. You'll see the difference in yourself.

Claudius Galen (131–201 CE), known simply as Galen, was a Greek anatomist, physician, and writer. He was the court physician to Emperor Marcus Aurelius of Rome from 160 CE to 200 CE.

Galen's major contribution was the collection and publication of what was then considered medical knowledge along with his own extensive research. In addition, he is especially noted for demonstrating that arteries carry blood and not air, and his knowledge and exploration of the brain, nervous system, spinal cord, and heart.

His theories formed the basis of European medicine until the Renaissance.

Andreas Vesalius (1514–1564), was a Flemish anatomist and surgeon. He is considered the father of modern anatomy. He was the private physician of the Holy Roman Emperor Charles V.

In 1543, Vesalius published a seven-volume text about the structure of the human body. The book contained the first accurate illustrations of internal human anatomy. The book caused a major controversy because it disproved many of the theories of Galen, then the leading medical authority. Unlike Galen, Vesalius studied the anatomy of the human body. He is known to be the first author of a comprehensive and systematic view of human anatomy.

Conclusion

The body is a complex machine. But properly taken care of, our bodies will function correctly so that we can have healthy and enjoyable lives.

The skeleton truly is the framework of the body. Muscles, from the largest to the smallest, are truly the workhorses. Working together, the two systems provide body structure, support, and movement.

The pitcher winds up and throws the ball. The batter swings. It's a long fly ball. You run after it and grab it just before it reaches the center field wall. Now that's a good example of your skeleton and muscles working together!

Glossary

bone
(BONE) hard tissue that provides support for the body (page 2)

cartilage
(KAR-tih-lij) smooth tissue that covers bone ends, ears, and nose (page 14)

cranium
(KRAY-nee-um) part of the skull that covers the brain (page 10)

disk
(DISK) soft tissue between two vertebrae (page 13)

joint
(JOINT) place where two bones come together (page 4)

ligament
(LIH-guh-ment) tough tissue connecting bone to bone (page 24)

marrow
(MAIR-oh) soft, spongy center of a bone (page 8)

muscle
(MUH-sul) body part that contracts and relaxes, producing movement (page 2)

skeleton
(SKEH-leh-tun) the body framework that supports and protects (page 2)

skull
(SKUL) hard, bony part of the skeleton consisting of the cranium and facial bones (page 7)

spinal column
(SPY-nul KAH-lum) central support bones of the skeleton (page 10)

tendon
(TEN-dun) cord-like tissue that connects muscle to bone (page 24)

vertebrae
(VER-teh-bray) bones that make up the spine (page 13)

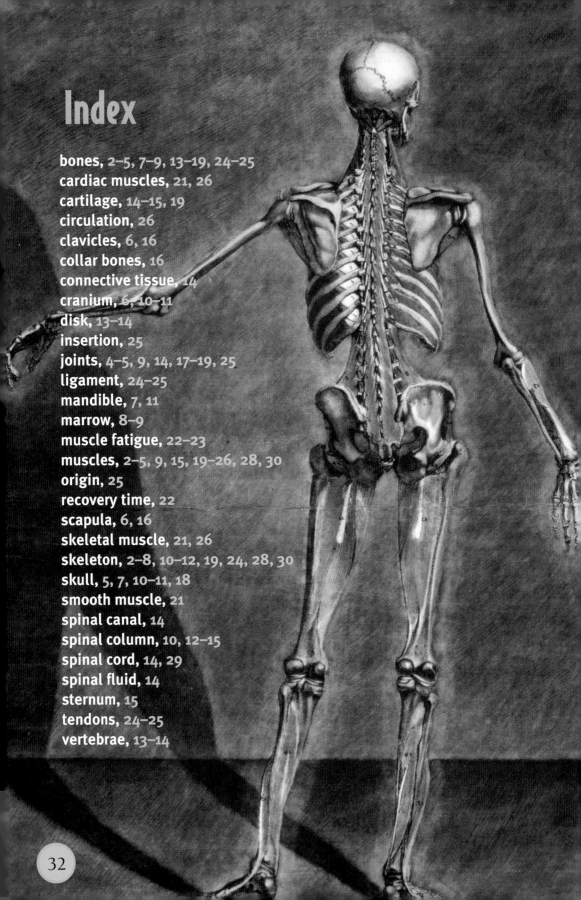

Index

Write in Your Science Journal

Choose one of the following prompts to write about in your journal. Make drawings, charts, or other graphic features to help you organize your thoughts.

1. Write about a way in which you take care of your bones and muscles. Think about your nutrition and exercise habits. (Make connections)

2. Review the text. Explain the differences and similarities among smooth, skeletal, and cardiac muscles. (Compare and contrast)

3. The author discusses how the skeletal and muscular systems work together. What other body systems do the skeletal and muscular systems interact with? (Make inferences)

Credits:
Executive Editor: Margaret McNamara
Art Director: Glenn Davis
Director of Photography: Lynn Shen
Literacy Consultants: Tammy Jones, Donna Schmeltekopf Clark
Concepting and Editorial Services: Logotek Limited

Photo Credits: Cover, Title Page, 6, 14A, 14B, 15A, 15B, 16, 19, 20, 23B, 24, 25, 28, 32: Gettyimages; Page 2: G. Schuster/zefa/Corbis Page 3: Duomo/Corbis; Page 4A: Richard T. Nowitz/Corbis; Page 4B: JEAN-PAUL PELISSIER/Reuters/Corbis; Page 8-9: Lester V. Bergman/Corbis; Page 9B: Jim Craigmyle/Corbis; Page 11A: Martin Philbey/ZUMA/Corbis; Page 11B: Bohemian Nomad Picturemakers/Corbis; Page 13: Mediscan/Corbis; Page 26: Biophoto Associates/Photo Researchers, Inc.; Page 27: Nucleus Medical Art/Visuals Unlimited, Inc.; Page 29A, 29B: Granger Collection

NAVIGATORS

BENCHMARK EDUCATION COMPANY
629 Fifth Avenue • Pelham, NY • 10803

Body Systems: Skeletal and Muscular

Do you know where your skeleton is? It's hiding beneath your muscles, of course, with all its bones. The skeleton is the foundation of the body. It gives us support, stability, protection, and shape. Without the skeleton, we would be little more than a puddle with eyes and a mouth.

ABOUT THE AUTHOR

Gary Rushworth has been a writer and poet for over thirty-five years. He completed his undergraduate work at Northeastern University and graduate studies at Towson University. He lives in New York with his wife, Vicki, and two Corgies named Hettie and Ellie. Gary and Vicki have three children, Mark, Sarah, and Merrell.

ISBN 978-1-4509-0734-7

9 781450 907347

Benchmark
EDUCATION

BENCHMARK EDUCATION COMPANY